5.00

Waste

Kay Davies
and
Wendy Oldfield

Starting Science

Books in the series

Animals
Floating and Sinking
Food
Sound and Music
Waste
Weather

About this book

This book takes a look at different types of waste and pollution in situations familiar to children. The children can relate their own experiences to the more general issues surrounding litter, man-made pollution, recycling and cleaning up waste, and nature's way of dealing with waste.

The activities and investigations are designed to be straightforward but fun, and flexible according to the abilities of the children. With the teacher's or parent's guidance they will be introduced to methods in scientific enquiry and recording. The children's involvement in this way should stimulate plenty of discussion.

The main picture and its commentary may be taken as a focal point for further discussion or as an introduction to the topic. Each chapter can form a basis for extended topic work.

Teachers will find that in using this book, they are reinforcing the other core subjects of language and mathematics. Through its topic approach *Waste* covers aspects of the National Science Curriculum for key stage 1 (levels 1 to 3), for the following Attainment Targets: Exploration of science (AT 1), The variety of life (AT 2), Human influences on the Earth (AT 5), and Types and uses of materials (AT 6).

First published in 1990 by
Wayland (Publishers) Ltd
61 Western Road, Hove
East Sussex, BN3 1JD, England

© Copyright 1990 Wayland (Publishers) Ltd

Typeset by Nicola Taylor, Wayland
Printed in Italy by Rotolito Lombarda
S.p.A., Milan
Bound in Belgium by Casterman S.A.

British Library Cataloguing in Publication Data
Davies, Kay 1946–
Waste.
1. Refuse
I. Title II. Oldfield, Wendy III. Series
628.4'4

ISBN 1 85210 999 8

Editor: Cally Chambers

CONTENTS

All the words that first appear in **bold** in the text or illustrations, are explained in the glossary.

The rubbish has been dumped at the tip. The seagulls can find scraps to eat.

A LOAD OF RUBBISH

Where do we put our **litter**? Look out for rubbish bins in the park, at school and at home.

Look in the litter bins in your classroom at the end of the day. See what people have thrown away.

Is everything you throw away really rubbish? Look at these things for instance. What could they be used for?

The plastic pot could be used to store marbles. Rubber bands could keep your pencils together.

WASTE PAPER

Collect things made of paper. Make two piles. One for those things we keep and one for those we throw away.

Our waste paper can be sent to a factory and made into clean new paper.

You can make your own new paper.

Put scraps of newspaper into a bucket with water. Mix it to a **pulp**. Make holes in the lid of a plastic box. Stretch a nylon stocking over the lid. Spread the paper pulp in a thin layer over the lid. Use a rolling-pin to flatten the pulp and squeeze the water out. When the pulp is dry, peel it off. Paint a picture on your new paper.

The wood from the trees will be made into paper.
Saving paper will help save the trees.

The **compost heap** is piled high with rotting leaves.

GOOD FOR THE GROUND

In autumn, leaves and fruit fall to the ground. These **rot** and become part of the soil. The goodness in them helps new plants to grow.

Rotting fruit becomes soft and brown. Green and white **mould** may grow on it.

Leave a banana skin and a bruised apple in a dish.

Draw the changes you see every day. Show how the bruise on your apple grows. Show how the banana skin changes colour. Draw in other changes you see.

Day	Apple	Banana skin
1.		
2.		
3.		

Carry on keeping your record over a few more days.

Wash your hands if you touch mouldy fruit.

Everyone is very hungry. They are enjoying their food.

LEFTOVERS

Look at all these scraps and wrappers left over in the kitchen.

Sort them into groups like this:

1. Things that will rot and be good for the ground.

2. Things that the birds might like to eat.

3. Things that will be thrown away.

Where can you put your groups of scraps and wrappers?

1. On the compost heap.

2. On the bird table.

3. In the dustbin.

Remember that even rubbish can be useful.

The tree stump is dead. But it makes a good home
for all sorts of animals and plants.

A LOT OF ROT

Wood that lies in damp, dark places begins to change. Small plants and animals, and rain all help to make the wood rot.

Look for a piece of rotting wood. Feel it and smell it. How is it different from new wood?

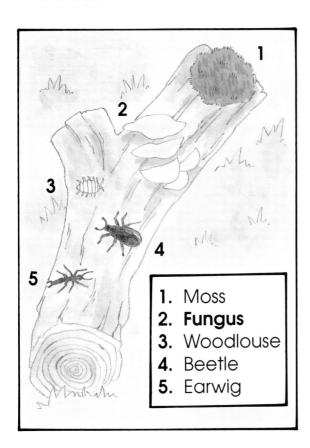

1. Moss
2. **Fungus**
3. Woodlouse
4. Beetle
5. Earwig

Here are some animals and plants that you might find on your log. Can you match up their names?

Draw your log. Draw some plants and animals on another piece of paper. Cut them out and stick them on your log.

Put the log back where you found it.

WATERWORKS

Every day we all use lots of water.

What sort of things do you use water for?

Before it reaches us it has to be cleaned. To do this it is passed through layers of stones and sand. These catch the dirt.

Put some water in a bucket. Drop garden soil, twigs, leaves and grass into the water.

Fill a funnel with clean **gravel** and hold it over an empty bucket. Pour some dirty water over the gravel. Can you see any difference in the water?

Do this lots of times to get your water clean.

The children are giving the dog a bath. The clean
water washes the dirt out of its hair.

There are no animals left in the pond. Rubbish has made the water dirty.

DEAD WATER

Animals cannot live in dirty water.

Try this in the summer. Put some dirty washing-up water into one bucket and some fresh clean water into another.

Leave them outside. Use a jar to take **samples**. Look at a new sample every week.

The dirty water begins to smell. No animals can live in this water.

The clean water is full of tiny creatures.

Fish and water-birds like to eat tiny creatures. So a clean pond can make a good home for them.

OIL CAN SPOIL

Find two dry fluffy feathers. Throw them in the air. Watch them float gently to the ground.

Dip one feather in some cooking oil. The feather looks sticky and heavy.

Throw the feathers in the air again. What do you notice now?

Birds with oil on their feathers find it hard to fly.

People sometimes rescue birds that have been covered in oil.

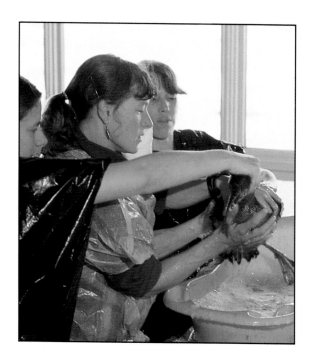

They try to clean the sticky mess from their feathers with lots of soapy water.

Oil from the shipwreck spills into the sea. It can kill sea birds when it sticks to their feathers.

WHAT A MESS!

Look out for litter on your way to school. Look in the park. Look in the shopping centre. Notice where the dirtiest place is.

Do you think it is nice to see litter? Where would you tidy it away to?

Litter makes our towns and countryside untidy. It can be dangerous.

Make a 'Keep Tidy' poster. Paint a big litter bin. Stick things you do not want on the bin.

Think of other ways to stop people dropping litter.

You could start a 'Tidy Up Litter' project in your school.

Everybody has had fun by the seaside. But look at all the litter they have left behind.

Factories and cars give out lots of smoke. It makes the air we breathe dirty. It can make us feel ill.

CLOUD OF SMOKE

How dirty is your air?

Put some light-coloured jelly into a jug. Pour in hot water to make it runny. Fill some shallow dishes with the jelly.

Leave your dishes in different places. Try a window sill, a path outside, on a wall and near a door.

Look at your dishes after two days.

You can see how much dirt and dust has collected on the jelly.

The dirtiest dish shows where there was most air **pollution**.

These·old pieces of metal are covered in **rust**. The rust has eaten away the shiny metal.

OLD IRON

Half fill a jar with water. Put a clean, shiny, iron nail in the water.

Look closely at your jar every day.

Soon you will notice that the nail begins to rust. The water turns brown too.

Try lots of metal things to see if they rust. Try things like bottle tops, bolts, paper clips, hair grips and safety pins. Use a different jar for each object.

Look for changes every day.

Only the things made of iron will rust.

SOMEBODY'S HOME

Some birds build their nests with dried grass, moss and twigs.

Sometimes they use pieces of string and plastic which we have thrown away.

Can you make a nest?

An old kettle provides a home for this robin and her babies.

Can you make a home out of rubbish? You could use cardboard boxes, sticks, old blankets, leaves and string.

The blackbird has made a nest for her babies. Can you see the bits of plastic she has used?

The model has been put together from all sorts of rubbish by an artist. Can you recognize anything that has been used to make it?

SOMETHING FROM NOTHING

You can make a model from throwaway things.

Try to make a caterpillar train like this.

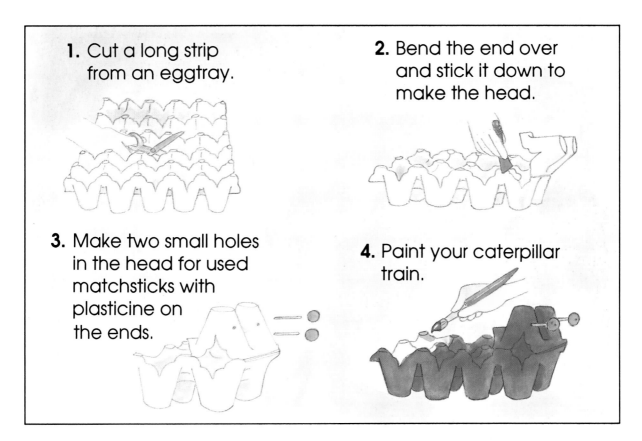

1. Cut a long strip from an eggtray.

2. Bend the end over and stick it down to make the head.

3. Make two small holes in the head for used matchsticks with plasticine on the ends.

4. Paint your caterpillar train.

Use the carriages to hold sweets, marbles and other things you want to keep tidy.

Try to make some more models from other kinds of boxes.

GLOSSARY

Compost heap A mixture of leaves and plant stalks that will rot.

Factory A place where machines are used to make things.

Fungus A living growth that prefers damp places. It is not an animal or a plant.

Gravel A mixture of small stones.

Litter Small pieces of rubbish, sometimes carelessly thrown away.

Mould A furry fungus that grows on rotting food.

Pollution Anything that spoils our soil, water or air.

Pulp Usually paper or wood that has been made wet and soft.

Rot When changes take place in living things, making them become soft and waste away.

Rust A red-brown covering on iron which has been wet or damp.

Samples Small amounts of something, showing what the rest is like.

PICTURE ACKNOWLEDGEMENTS

Bruce Coleman Ltd. 4 (Price),12 (Krasemann); Chapel Studios (Zul Mukhida) cover, 5, 6,9,13,14,18 top, 20, 21, 23 both, 25 both, 26 top,29; Eye Ubiquitous 11 (Seheult), 28 (Parkin); Eric and David Hosking 7; Frank Lane Picture Agency 26 bottom (Withers), 27 (Wilmshurst); Oxford Scientific Films 16 (Chillmaid); Papilio 8, 24; Photri 19; Rex Features 18 bottom; Tim Woodcock 10; ZEFA 15, 22.
Artwork illustrations by Rebecca Archer.
The publishers would also like to thank St. Andrews C. E. School, Hove, East Sussex, for their kind co-operation.

FINDING OUT MORE

Books to read:

Lots of Rot by Vicki Cobb (A & C Black, 1988)
Scraps of Wraps by Vicki Cobb (A & C Black, 1988)

The following groups will be able to help you find out more about waste and pollution.

Council for Environmental
   Education
School of Education
University of Reading
London Road
Reading   RG1 5AQ

Friends of the Earth (UK)
Education Office
26-28 Underwood Street
London
N1 7JQ

Greenpeace (UK)
30-31 Islington Green
London   N1 8XE

National Society for Clean Air
136 North Street
Brighton   BN1 1RG

Reactivart (Children's
   environmental workshops)
74 Chelsea Park Gardens
London
SW3 6AE

Royal Society for Nature
   Conservation (Watch)
22 The Green
Nettleham
Lincoln
LN2 2NR

Tidy Britain Group
Education Office
The Pier
Wigan
WN3 4EX

World Wide Fund for Nature
Panda House
Weyside Park
Catteshall Lane
Godalming
Surrey
GU7 1XR

INDEX